Lockheed-Martin
F-22 Raptor

Mike Wallace & Bill Holder

Schiffer Military/Aviation History
Atglen, PA

Acknowledgments

The authors wish to thank the following individuals and organizations for their help in this book:

1. Bobby Mixon, Aeronautical Systems Center Office of Information, USAF, Wright-Patterson AFB, Ohio

2. Richard Kennedy, Jim Stump, General Electric Company, Cincinnati, Ohio

3. Mike Tull, Public Affairs Office, Boeing Company, Seattle, Washington.

4. Paul Ferguson, Aeronautical Systems Center History Office, Wright-Patterson AFB, Ohio.

Book Design by Ian Robertson.

Copyright © 1998 by Mike Wallace & Bill Holder.
Library of Congress Catalog Number: 98-84030

All rights reserved. No part of this work may be reproduced or used in any forms or by any means – graphic, electronic or mechanical, including photocopying or information storage and retrieval systems – without written permission from the copyright holder.

Printed in China.
ISBN: 0-7643-0558-1

We are interested in hearing from authors with book ideas on related topics.

Published by Schiffer Publishing Ltd.
77 Lower Valley Road
Atglen, PA 19310
Phone: (610) 593-1777
FAX: (610) 593-2002
E-mail: schifferbk@aol.com
Please write for a free catalog.
This book may be purchased from the publisher.
Please include $3.95 postage.
Try your bookstore first.

Table of Contents

Chapter 1	Determining the Need for the F-22	4
Chapter 2	The Competition for the Contract	12
Chapter 3	The YF-23 and F120 Engine	16
Chapter 4	Other Versions of the F-22	26
Chapter 5	The F-22, Parts and Pieces	30
Chapter 6	The F-22 Ground/Flight Test Program	44
Chapter 7	F-22 Production and Operational Service	58
	Color Gallery	70

Chapter 1:
Determining the Need for the F-22

April 9, 1997, saw the F-22 Raptor production fighter "rolled out" for public view. In a ceremony attended by 3,000 guests including the Secretary of the Air Force, Sheila Widnall, and the Air Force Chief of Staff, General Ronald Fogleman, the Raptor was described as "air dominant," meaning that it could completely overwhelm any other fighter, current or planned.

The gist of the many speeches was that the F-22 would never offer a "fair" fight; instead, on the offensive and usually undetected, the F-22 would simply win *any* air-to-air engagement. The significance of this is that the F-22 will dominate the skies and provide a formidable deterrent to any nation threatening U.S. interests.

As with many high-tech Department of Defense programs, the F-22 is expensive—the Air Force estimates more than $90 million per aircraft. However, the cost of not fielding the plane could be much higher. According to a statement by the F-22 contractor team, Lockheed-Boeing-Pratt & Whitney: "No U.S. ground soldier has been killed by enemy air power in 40 years." One goal of the program is to continue this record.

Perhaps even more importantly, in view of the world's instability and the capabilities of foreign-owned aircraft, the Raptor will discourage belligerence. It's important to note that national allegiances change and it's possible that U.S. forces in the future might face advanced Russian, French, British, Japanese, Indian, Swedish and even U.S. fighters, as well as a wide variety of surface-to-air and air-to-air missiles. However, the F-22 might preclude conflicts by giving the U.S. the capability to establish and maintain air dominance in any situation vital to the U.S. Although such claims might seem outlandish, the Raptor stands ready to live up to them. Just how it can do this is based upon six major factors.

Stealth, or low observables technology. The Raptor has an extremely small radar cross-section which makes it nearly invisible to enemy radar. This allows the F-22 to penetrate the adversary's airspace because, by the time an enemy radar system detects the Raptor, it's too late to do anything about it.

In addition, airborne radar have both electrical power limits and antenna size limits which make detecting and tracking the stealthy F-22 nearly impossible.

Stealth technology allows the F-22 pilot more flexibility in his tactics than does any other fighter aircraft. Armed with

WHEN THE FIRST F-22 ROLLS OFF THE ASSEMBLY LINE IT WILL SEND AN UNMISTAKABLE SIGNAL TO THE REST OF THE WORLD. AMERICA WILL NEVER RELINQUISH OUR DOMINANCE OF THE SKY. THAT SAME SIGNAL WAS HEARD LOUD AND CLEAR IN THE GULF WAR. SIMPLY BY FLEXING OUR MUSCLES, AMERICA FORCED THE IRAQI AIR FORCE TO TAKE COVER. AS A RESULT, THE WAR WAS SHORTENED AND ALLIED LIVES WERE SAVED. BUT THE AIR SUPERIORITY FIGHTER THAT FLEW IN IRAQ WILL BE 30 YEARS OLD BY THE TIME THE FIRST F-22 SQUADRON IS READY. WHICH IS WHY AMERICA NEEDS THE F-22 AS MUCH AS POTENTIAL ENEMIES FEAR IT.

It can win a war, or prevent one.

The descriptor most heard about the F-22 was "Dominance." USAF officials at the official roll-out in April 1997 stated, "We don't want a fair fight!" (Lockheed Photo)

The Lockheed P-38 fighter was a dominant aircraft during World War II. The same company and USAF hope the same will be true for the F-22. (Lockheed Advertisement)

The April 9, 1997 rollout of the F-22 received international attention with about 3,000 dignitaries including Air Force and industry personnel present. (Air Force Photo)

internally-carried air-to-air weapons, the Raptor will fly undetected, destroy enemy aircraft and dominate tactical and strategic airspace.

Speed. Traditionally, fighter pilots never could have enough speed. The F-22's top speed is classified but believed to be in excess of Mach 2.5, or faster than 1800 mph.

What distinguishes the F-22 from almost all other fighters is its ability to *supercruise*, or fly supersonically without using afterburners. In a jet engine, the afterburner power comes from squirting fuel into the "controlled explosion" known as the exhaust. The results: a great boost in power which can put most modern fighters at supersonic speeds but at the expense of dramatic fuel consumption. The F-22, however, with its pair of Pratt & Whitney F119 engines, can fly supersonically without afterburner for relatively long periods of time.

Just how long the Raptor can supercruise is classified, but certainly long enough to give it enormous tactical advantages such as only short exposure to surface-to-air missile systems, reduced time to close on enemy aircraft, ability to cover great distances in short times allowing more engagements or interceptions, and greater survivability since the pilot can retreat quickly and have enough fuel to get home.

Range. Another classified number, but the Air Force claims a subsonic range greatly surpassing that of the F-15 Eagle and a supersonic range two-and-a-half times that of the F-15. This greatly-increased range means fewer in-flight refuelings and increased basing flexibility—which complicates any enemy's attack plans.

Greater range also allows the Raptor to control more airspace or spend more time in the immediate battle area. These

This Lockheed advertisement puts forth some of the strong arguments for the F-22. The fact that other fighters are being built around the world, and the US wants to continue its lead in air superiority. (Lockheed Advertisement)

Currently, the F/A-18C/D is the mainstay of the Navy/Marine air forces. Unlike the Air Force, the Navy continued to modify the F/A-18 as opposed to a completely-new aircraft. (US Navy Photo)

Introduction of the YF-22 prototype. Note the NATF lettering behind the plane, a Navy version of the aircraft was considered early in the program. (USAF Photo)

F/A-18E/F HORNET
NEXT GENERATION NIGHT STRIKE FIGHTER

As of press time, the F/A-18E/F, shown here in artist's concept, was a main competitor for Department of Defense dollars. The fact that it was flying sooner, and is considerably less expensive than the F-22, makes it formidable competition. However, in late 1997, the F/A-18E/F was having wing flutter problems. (McDonnell Douglas Photo)

factors make it especially effective in protecting U.S. interests in regional conflicts.

Agility. Although difficult to define, the degree of agility—or maneuverability—is the set of flying qualities which allow one fighter to dominate another in a close-in dogfight. With its vectored thrust engines (in which the exhaust nozzles can be directed up or down 20 degrees from level) which give the Raptor nearly unprecedented maneuverability, and its integrated avionics, the F-22 will certainly have an enormous edge in close-in combat.

It should be stated that the purpose of the F-22 is not dogfighting but rather deadly efficiency from a distance. But, if the enemy somehow closes the distance, the Raptor's agility will bring about its destruction.

Integrated Avionics. Touched upon earlier in this text, integrated avionics means that the F-22's sensors, navigation equipment, communications gear, electronic countermeasures, and other systems will be linked together by high-speed processors and software. This linkage and processing speed gives the pilot vastly improved situational awareness which, when combined with stealth, aircraft speed, and weaponry, make the F-22 extremely lethal.

Raptor's avionics features Very High Speed Integrated Circuits (VHSIC), a design featuring common modules allowing rapid replacement, fiber optics for the rapid transfer of high-volume data, and military-standard Ada language software. The modular design allows for long-term, cost-effective upgrading with new technologies and systems.

The first F-22 is being towed to its position in the rollout ceremony. This photo illustrates some of the complex angles on the forward section of the F-22. These angles result in a very small radar cross section which makes this fighter stealthy. (Air Force Photo)

Air-to-Ground Capability. The technologies enabling the F-22 to be the most lethal air-dominant fighter in the world also make it inherently deadly in the air-to-ground role. Current plans are for the Raptor to eventually replace the F-117 Nighthawk—the stealthy "star" of Desert Storm—for carrying out certain category, precision bombing missions.

The F-22 will be able to carry at least two 1000 pound, Global Positioning System-guided, Joint Direct Attack Munitions (JDAM) in any kind of weather for the purposes of strategic attack, interdiction, or suppression of enemy air defenses.

Other Enhancements
There are two other factors which will contribute to the effectiveness of the F-22 fleet. One is increased reliability and maintainability, R&M; the other is reduced airlift.

Increased R&M was a design goal for the Raptor. Each component was tested for ease of maintenance in the field as well as producibility and durability. This means that things won't break as often, but if they do, they can be replaced quickly.

The hoped-for result: an F-22 will be able to conduct 8.5 sorties before requiring major maintenance. This compares favorably with the F-15's 5.4 sorties. And increased R&M will be capable of sustained sortie rates much higher than those of the F-15 with much less logistical support and lower life-cycle costs.

Reduced airlift means that the Air Force will need far fewer people and equipment to repair and maintain the Raptor. One

The Air Force's F-15 Eagle, the dominant air superiority fighter since the 1970s, will be surpassed in capability by the F-22. No F-15 has ever been lost to enemy fire, a fact worth noting in view of Desert Storm in the early 1990s. (McDonnell Douglas Photo)

The new and the old, the YF-22 prototype (after just firing an AIM-9 missile) and an F-15 Eagle. (USAF Photo)

estimate is 50 percent fewer airlift and 40 percent fewer people than an F-15 squadron. The F-22 will be able to deploy without an avionics intermediate shop and with 75 percent fewer hydraulic carts, generators, air conditioning units, and other items needed for repairs and maintenance.

These numbers will become more and more important as the U.S. presence overseas declines in the near future.

As was stated at the beginning: the F-22 won't offer any adversary a fair fight. Telling the story of how this ultra-lethal package came about is the aim of this book. The success of this fighter will be measured hopefully not in combat but rather in rugged flight testing through 2004 and beyond.

Similar to the US Navy transition to the F/A-18E/F, the Air Force greatly modified the F-15C/D (known as the F-15E), with the addition of much new technology-i.e. the LANTIRN Pods seen under the wings which enhanced the ground attack capability. (McDonnell Douglas Photo)

Chapter 2:
The Competition for the Contract

What came to be known as the F-22 Raptor evolved from a lengthy and carefully-tailored program in the early 1980s known as the Advanced Tactical Fighter, or just ATF.

Competition was the cornerstone of the ATF program. There was competition with the Soviet Union whose military, according to intelligence reports, was on the verge of creating fighter aircraft equal to or even superior to the F-15 Eagle. The Soviets seemed to iterate a new type of MiG or Sukhoi fighter every five years or so—and each one more capable than the last.

The Air Force was concerned for its F-15s which would be aged Eagles by the 1990s, but a new fighter program faced competition for dollars from not only other Air Force programs but also programs of the other military branches. The Department of Defense had a finite number of dollars and the Secretary of Defense had to select from among numerous proposals the ones which promised the greatest military advantage in the near future. To add a fighter program would take careful planning, exploitation of maturing technologies, and presenting clear-cut arguments to Congress.

Complicating the task was the fact that choices were among "apples and oranges." For example, which was needed more: a new Air Force fighter or a new Navy aircraft carrier? Or could the nation afford both? It was up to the Secretary of Defense—with Congressional approval—to decide.

Defining the problem of replacing the F-15 in the future and offering the best solution was the very crucial task given to the Air Force. An organization at Wright-Patterson Air Force Base, Ohio, known as Aeronautical Systems Division's Deputy for Development Planning, gathered information and weighed options.

Development Planning forged ahead in conjunction with other Air Force offices including: the Tactical Air Command which emphasized their need for a more advanced fighter aircraft for the future; the Foreign Technology Division which analyzed foreign military advancements and potential threats against the U.S.; and Air Force laboratories which, through both in-house work and contracted research, "marketed" promising aerospace technologies which might become part of a new fighter.

This Grumman forward-swept-wing concept artist concept was released by the company in the mid-1980s and could have represented thinking in the ATF competition. (Grumman Drawing)

A Boeing design which featured swing wings and engine inlets on top of the fuselage reducing its infrared signature. This concept was released during the initial ATF competition time period. (Boeing Drawing)

A futuristic gull-wing concept was released by Boeing. One could only speculate on the type of materials that would have been required for this concept. Note that the engines were mounted underneath the curved sections of the wing. (Boeing Drawing)

Though it was pre-ATF concept, this Boeing drawing shows vectored thrust nozzles which would eventually be incorporated into the final F-22 design. Note the tiny control canards located on the lower portion of the front fuselage. (Boeing Drawing)

Development Planning's director, Stan Tremaine, began touting the Advanced Tactical Fighter as "...insurance for the 1990s and beyond."

Air Force colonel, Albert C. Piccirillo, took the program's reins in early 1983. His first challenge was to take the information from the aforementioned Air Force agencies, arrive at statements of the problems that a fighter of the 1990's and beyond might face and offer defense contractors a chance to write out their "solutions."

To do this, the Air Force sent out Request for Proposals (RFPs) to each of the major airplane builders as well as to the two fighter engine manufacturers. In September 1983, after exhaustively evaluating the proposals, Aeronautical Systems Division (ASD) awarded airframe contracts to Boeing, General Dynamics, Grumman, Lockheed, McDonnell Douglas, Northrop and Rockwell. Similar contracts went to Pratt & Whitney and General Electric.

Each company was to present its respective design concepts to the Air Force by the following spring. ASD required that the conceptual designs for the airframe include performance against all anticipated threats—including weapons, electronic jamming, cost, risk and supportability.

Supportability received special emphasis because of the Air Force's need to operate from "austere or battle-damaged sites" as well as the recognition of supportability problems which plagued previous programs.

Colonel Piccirillo explained, "The contractors must integrate the best features of current technologies. Some areas to be considered (would) be new materials, such as composites and advanced metallics; new electronics, including advanced cockpit automation, integrated fire and flight controls, advanced radars and sensors; vectored thrust; built-in test and support equipment; and low observables technology.

"The ATF (Advanced Tactical Fighter) will be the sum of the proper integration of all these things into a blended weapon system that will achieve optimum air superiority."

The engine contractors were given similar requirements. Supportability and maintainability were coupled with very advanced performance parameters—especially those for "supercruise," or supersonic speed without using afterburners.

Ease of maintainability—in the real-world, blue-suit technician sense—was of especially high importance. Parked on the ramp, a fighter aircraft simply is an expensive target; airborne,

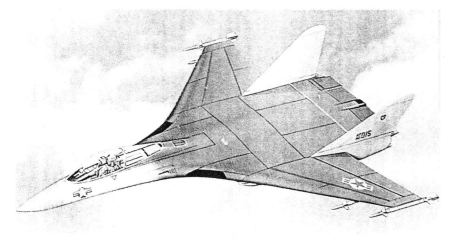

This Rockwell drawing was published as ATF candidate configuration. A huge aircraft, the plane proposed over a thousand square feet of wing area. (Rockwell Drawing)

The Competition for the Contract   13

This ceremony at Lockheed celebrated the rollout of the YF-22 prototype. In the same time period, a similar ceremony took place at Northrop with its YF-23 prototype. This lowered the seven initial proposals from all the major airframe contractors down to two. (USAF Photo)

it is a formidable protector. Knowing this, the ATF management team had Pratt & Whitney and GE engine designers go to Air Force flight lines and pull engines out of F-15s and F-16s to learn firsthand what the maintenance people faced in the field.

An example of the obvious: why were there so many different-sized bolts? If designers could reduce this number, maintenance time would decrease.

Involving—as early on in the program as possible—designers, engineers, maintenance people, systems safety people, and others representing as many aerospace disciplines and technologies as possible became standard operating procedure for the ATF. It also became the model for other programs to follow since the Air Force believed that this approach would ultimately save money over the life of any program.

Following a six-month evaluation process, the Air Force selected two contractors, Lockheed and Northrop, to continue with their ATF plans. Contracts, each worth $691 million, were awarded on October 31, 1986.

Contracts marked the beginning of what the Air Force called the "demonstration/validation phase," a 50-month effort during which the contractors would demonstrate their airframe designs and avionics concepts prior to full-scale development scheduled, at that time, to start in 1991.

There were, and continue to be, high hopes for the ATF. The then-commander of Aeronautical Systems Division, Lieutenant General William E. Thurman, said, "The ATF will be the Air Force's air superiority fighter for the year 2000 and beyond. The aeronautical world once again will focus on Dayton, Ohio, and Wright Field as we continue work to create the Air Force of tomorrow."

The demonstration/validation phase included building flying prototypes: Lockheed's YF-22A and Northrop's YF-23A. Each contractor was to build two aircraft: one for each of the two prototype engines: Pratt & Whitney's YF119 and GE's YF120.

General Thurman added, "The flying prototypes will be a major part of the program. We need flight testing to demonstrate the proper balance among the critical ATF characteristics of supersonic cruise, high maneuverability, and low observable radar signatures as well as excellent fighter handling qualities prior to pinning down the designs for full-scale development and production.

"But the flying prototypes will be no more important than the avionics prototypes to be developed. The avionics prototypes, or mock-ups of the ATF's integrated electronic systems, at first will be ground-based, and they'll take into consideration the insertion of new technologies such as VHSIC (Very High Speed Integrated Circuits). With proper planning,

For the ATF competition, there were seven airframes, but only two engine proposals. Here, the GE F120 is shown in test, the engine which would eventually lose to the P&W F119. (GE Photo)

we'll be able to pull together the various avionics program elements and supporting technologies into a form suitable for full-scale ATF development."

At the time of demonstration/validation contract awards, the Air Force planned a buy of 750 aircraft. Also at that time, an Air Force/Navy agreement, the Navy began evaluating the ATF as a potential replacement for the F-14 fleet air defense interceptor for the turn of the century; the Air Force began looking at the Navy's heavily-classified A-X program for possible ground-based use. The Navy estimated their need at 550 naval variants of the ATF. Of course, that would never come to pass.

Chapter 3:
The YF-23 and F120 Engine

Granted, this book is devoted to the story of the F-22, but a major part of that story comes with the aircraft that it had to defeat. The YF-23, which was a joint Northrop/McDonnell effort, was a worthy adversary to be sure and certainly is a big part of the story. Much of the technology in that plane could well find itself manifested in future fighter aircraft.

The other major story here is of the major engine system that was not selected for the program—the General Electric F120 engine.

Some of the criteria that were listed as reasons for the winning entry was that the F-22/F119 combination would probably cost less in the long run. With the excellence displayed by both the airframes and engines, it was a gut-wrenching decision for the Air Force to make. Its consequence was so lofty because of the huge dollars that would be invested in the program through the years.

Let's take a look at the YF-23 and F120 which could well have combined to produce an outstanding fighter weapon system.

The YF-23

Right off the bat, it has to be definitely stated that the YF-23 sure didn't look like anything that the US Air Force had ever flown. The overall look was something straight out of Star Wars, but there was a technical reason for everything that looked so strange. The winning F-22, though, appeared more of the expected design, although stretching the point in several aspects.

Quite frankly, the YF-23 had a flat look about it including the way the engines were mounted. The engine inlets were located under the wings with the engines flared into the wings. Everything about the body spoke of speed and stealth, appearing to be able to glide through the atmosphere with practically no drag. Even the rear tail surfaces seemed to adopt the arrow-shape theory being bent close to the horizontal.

It was a remarkable aircraft, and at the time, was given an excellent chance of winning the competition. Let's check out the "hows and whys" of this futuristic concept and see why many think that some of its technologies and concepts

The YF-22(left) and YF-23 sitting on the tarmac at Edwards Air Force Base where they competed in the Advanced Tactical Fighter(ATF) competition. (USAF Photo)

This airborne photo of the two ATF prototypes, the YF-23 on the left, shows the similarities as well as differences in two planes. Note the vectored thrust of the YF-22 and the diamond wing configuration of the YF-23. (USAF Photo)

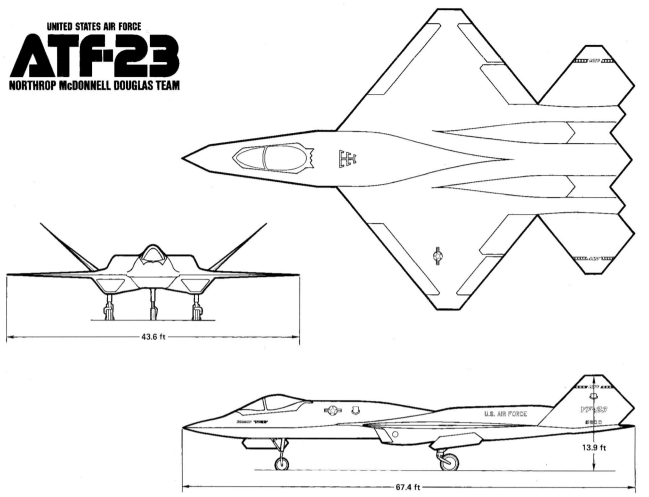

The dimensions of the YF-23 are shown by this Northrop drawing. The plane is very similar in size to the YF-22. (Northrop Photo)

could well appear in the fighters of the future. Note that the descriptions are all in the present tense because, in the late 1990s, both YF-23 prototypes are still in existence and should continue to be around for many years. They have found a new mission with NASA as testbed aircraft performing scientific missions.

That unique shape again was the big difference with the YF-23. The unique lines and curves of the design could best be described as the blending of aerodynamics and low-observable shaping. And even though the design might appear unconventional by today's standards, don't be surprised if it doesn't appear sometime in the future.

With its earlier, highly-unorthodox B-2 Bomber design, it was maybe not that surprising that the company might repeat the practice with this effort.

There are some remote similarities of the design with the SR-71 giving the same bat-like appearance. The shape tends to disguise the actual size of the plane, but the YF-23's fuselage is actually longer than an F-15. In actual dimensions,

The official logo for the YF-23 identifying the three major players in the aborted program. (Northrop Photo)

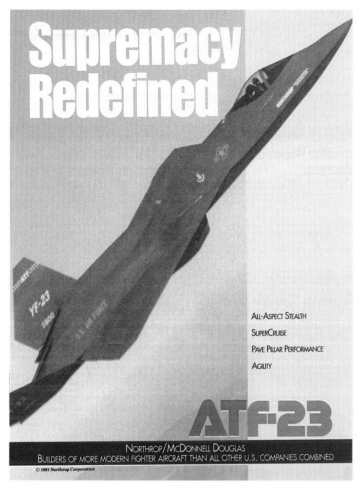

the YF-23 has a fuselage length of 67.4 feet and a 43.6 foot wingspan. It's probably those gently-sloping tail surfaces that give the illusion of greater length. Those surfaces, by the way, are attached at a single pivot point.

The YF-23 was definitely a flatter design that the competitor being only 14 feet in height compared to the more conventional F-22 at a lofty 18 feet. Its wingspan is about 43 feet, almost exactly that of the F-22.

Interestingly, the performance of the planes was almost identical with the maximum speed capability at about Mach 2.2 and a cruising speed of about Mach 1.6.

Looking down on the topside, there seems to be a clipped triangular look about the YF-23. The wings are very close to being equilateral triangles. The leading edges project a swept wing look, but the trailing edge has forward sweep. Go figure.

On the rear of the aircraft from above, there is an interesting pattern on the rear surfaces and the aft fuselage. Actually what is created is a sawtooth pattern which was so much a part of the super stealthy F-117A. Coincidence? Probably not, because the same shaping is observed on the panels of the wing surface.

The bubble canopy is the high point of the fuselage with the area behind the cockpit continuing with an increased diameter until it drops down about mid-wing.

Unlike the current generation of US fighters, the engine inlets are NOT flushed up close to the fuselage. Instead, they hang separately from the bottom of the wings and sport a trapezoidal shape from the front viewing position.

The engines, though, are buried deep in the aft fuselage, well rear of the cockpit. The engine ducts simultaneously

This 1991 Northrop advertisement flaunts the YF-23. (Northrop Advertisement)

The flatness of the fuselage and the extreme outward cant of the rear control surfaces is very evident in this head-on shot of the YF-23. (USAF Photo)

The YF-23 and F-120 Engine

Unlike any other modern fighter, this YF-23 shows its aerodynamic control surfaces, both on the forward and trailing edges on the wings. (USAF Photo)

Note that the cockpit is resting on a hump providing excellent visibility in all directions. (USAF Photo)

Stealthiness is enhanced by the blending of the wing-fuselage blending which all seems to run together. (USAF Photo)

sweep upwards and inboard, effectively shielding the engine compressor from head-on radar pulses.

On the other end of the engine, there is considerable protection against the searing heat. The material surrounding the exhaust ducts is capable of providing protection up to 1000 degrees F. As a result, the underside of the craft, only a couple inches away, will never get any hotter than 280 degrees, making the YF-23 very difficult to detect by enemy infrared detection systems.

The cockpit features a fixed vertical bar, slightly forward of the mid-point of the cockpit, a definite shift from the single-piece bubble canopies of current systems. Like we said, there is very little with the -23 that is like anything else.

The front of the fuselage comes to a distinct edge which reaches back to the leading edge of the wings. The contractor called it a 'chine' with its stated purpose being for low observable characteristics and aerodynamics concerns. Also located on the forward fuselage were flush-mounted air data sensors which could operate at both supersonic and subsonic speeds. The refueling probe is located far back on the top of the fuselage.

There are several weapon bay doors on the underneath side as the planned AIM-9 Sidewinder and AMRAAM missiles would have been carried internally. Hanging any external ordnance on this sleek machine would certainly have negated much of the aerodynamic advantages it enjoys.

The YF-23 carries a sophisticated Vehicle Management System (VMS) that monitored the plane's many onboard systems. The primary purpose of the VMS was to allow the aircraft to make maximum use of all its flight surfaces. With the

The stealth aspects of the YF-23 are particularly evident where the plane offers no flat surfaces to radar beams. Note the sharp edge which separates the upper and lower surfaces. Also note the thinness of the rear vertical control surfaces. (USAF Photo)

Note that the YF-23 wings incorporate both forward and rear sweep, probably the only modern fighter to do so. (USAF Photo)

The YF-23 engine inlets are located on top of the fuselage for protection against infrared detection. Also, the zig-zag pattern on the rear of the plane deflects radar signals away from the radar transmitter. Similar patterns are visible on the F-117 and B-2. (USAF Photo)

There are no flaps on the rear control surfaces. The whole surface pivots on a single pin for maximum maneuverability. (USAF Photo)

proper software in place, leading and trailing edge flaps and the two large tail surfaces can be moved for improved stability and maneuvering.

The VMS can also optimize the functioning of the aircraft hydraulic system, thus reducing the overall demands of the system allowing lighter components to be used. Should a faulty component come to light in the hydraulic system, the VMS will locate it, isolate it, and reroute the remaining fluid.

The performance requirements for the YF-23 were extremely challenging, and during the flight test program, it accomplished many of them. The plane's goal was high survivability, supersonic cruise capabilities, low-observable technologies. Also, the plane was designed to operate with less than half the maintenance required for current fighters and be more than twice as reliable, with a turnaround time of less than half than required by present systems.

With its unique shaping, the YF-23 proved to be an excellent performer at high angles of attack (AOA). But the -23 didn't prove to be as maneuverable mostly because it didn't have the thrust vector control system carried by the F-22. That innovation enables the F-22 to turn at very low and very high speeds.

It really came down to comparing the superior stealth capabilities of the YF-23 versus the better maneuvering capability of the -22. The decision was made to go with the latter concept. Would the YF-23 have made a good ATF? Undoubtedly.

Needless to say, there was a large YF-23 production team organized for the hoped-for contract. The team was comprised of thousands of workers in companies representing 31 states, all of whom were certainly unhappy the day the F-22 winning decision was announced.

So what happened to the YF-23 prototypes after the sad decision came down? Fortunately, the advanced technology of the aircraft would be able to serve the US aerospace industry in another style, that of an research and development tool.

This YF-23 prepares to refuel from a KC-135 tanker during the ATF competition. (USAF Photo)

The YF-23 and F-120 Engine

This YF-23 gets refueled while the winning YF-22 awaits its turn. (USAF Photo)

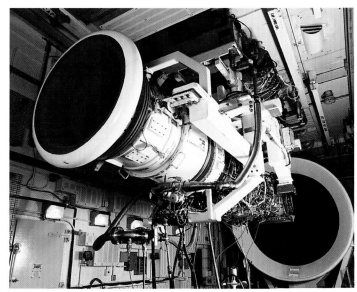

Testing was elaborate for the General Electric F120 variable-cycle engine shown here on a company test stand. (General Electric Photo)

The National Aeronautics and Space Administration used one of the prototypes to study aircraft loadings. The purpose of the study was to improve the accuracy of current strain gauge loads calibration methods. The YF-23 was selected for the study because of the composite materials used in the -23's structure.

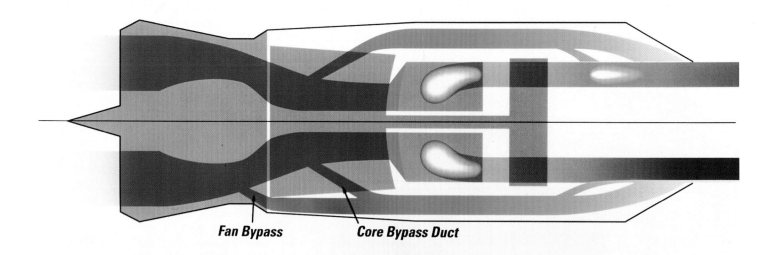

The two modes of operation of the F120 engine are shown by this illustration. Although this engine lost the ATF competition, it could find its way into future fighters. (General Electric Photo)

The F120 Engine

The fact that this engine was the loser certainly wasn't a knock at this excellent powerplant. It was given a one-on-one test with one of each prototype ATF model carrying a pair of the GE engines.

Was the engine operating at a disadvantage because of the just-mentioned TVC system possessed by the F-22's P&W F119 engine? That argument certainly can be offered since that arrangement just wasn't possible with the rear fuselage design of the YF-23.

To keep things pretty equal, though, the F120 had the same 35,000 pound thrust capability of the Pratt & Whitney model.

The F120 is a variable cycle technology engine which enabled it to operate like a conventional turbojet at supersonic speed, while demonstrating the characteristics of a more fuel-efficient turbofan at subsonic cruise speeds.

The main components of the high-tech engine system include a two-dimensional, vectoring exhaust nozzle, and a dual-spool, vaneless, counter-rotating turbine. Also, a number of improved materials found their way into the design, along with the fact that there were some 40 percent fewer parts than GE's F120 fighter engine.

Testing on the engine was highly successful with a record for test duration and parameters measured at the Air Force's Arnold Engineering Development Center. The 37 hour test period was the longest continuous engine test in the history of the facility, a period which saw some 875,000 measurements taken.

The 35,000 pound thrust class F120 engine shown undergoing testing attached to the so-called Turbulence Control Structure. (General Electric Photo)

Chapter 4:
Other Versions of the F-22

The story is old, going back through a number of previous aircraft developments. The reasoning behind having a common aircraft for both the Air Force and Navy is obvious— It's called economics. But for various reasons, it just hasn't proved successful as the requirements of each service dictate different design parameters making the common design very difficult.

It was tried with the F-111 in the 1960s with the F-111A being the Air Force version, while the Navy version was to be called the F-111B. As is well known, the F-111A would evolve into the potent FB-111 bomber and other versions which would serve into the 1990s in several different capacities. After a disastrous flight test program, the Navy version was shelved for good.

Later, there was an excellent example of the concept, but it didn't quite happen by the above plan. During the 1970s, the Navy was proving quite vividly the effectiveness of its F-4 Phantom fighter. It was then decided by the Secretary of Defense to force the Air Force to accept the aircraft into its fold. It did so grudgingly, but later learned to love the brutish fighter.

Then came the Lightweight Fighter Competition in the early 1970s, with the winning design hopefully being accepted by both services. The General Dynamics F-16 was a light and agile machine that was selected by the Air Force, but not the Navy, which went to its historical stance of requiring a two-engine aircraft. The competing Northrop F-17 fulfilled that requirement, a model which would eventually evolve into the F/A-18 Hornet which has seen three different versions, the latest E/F model which will be the prime Navy fighter well into the next century.

Then came the subject of this book, the F-22, with the same type of dual-service considerations. As with the previous examples, it didn't happen here either. Possibly the pres-

Devised as the answer to both Navy and Air Force light bomber requirements, the F-111 was devised as a common weapon. It turned out to be feasible for a number of reasons. Like the F-22, it never came to pass. This particular F-111 in the early 1980s was known as the Air Force's Tactical Jamming System. (USAF Photo)

But alas, it didn't happen even though there were official studies carried out, looking at Naval versions of both the losing F-23 Northrop-McDonnell Douglas entry as well as the winning F-22.

In the early 1990s, Northrop and McDonnell Douglas engineers looked at their swoopy YF-23 design and how it could fit into an aircraft carrier environment. Coined the NATF, the Naval mission was seen as an air superiority fighter mission with a significant strike capability.

In fact, it all began in March of 1986 when both the Navy and Air Force evaluated both designs. It was obvious that if the Navy could accept a version of one of the designs, significant savings could be realized in engine, avionics, materials and manufacturing techniques. The major differences in the two service versions, had the concept come to fruition, involved certain structural components, such as the landing gear, which would have to be beefier to withstand the catapult takeoffs and arrested landings aboard carriers. The Navy version would also have had to have low-speed flying qualities for carrier operations.

Once the F-22 design was selected, the Air Force insisted that it could fulfill the Navy mission requirements. Industry officials stated that a modified F-22 could replace the Navy fleet of A-6 attack aircraft as well as the F-14 fleet. It was argued that a Navy F-22 could also eliminate the need for an A-6 replacement aircraft, a plan that was being seriously considered at the time.

During the same time period, the Navy was looking in a different direction with its so-called AX program. Guess it's not surprising that a Lockheed-led team proposed an AX design which looked amazingly like the F-22. The differences were in the wing design, with the "F-22 AX" featuring a swing-wing for carrier deployment, a larger radar for increased range, along with unique subsystems for special naval operations and a lengthened fuselage. It, however, could well have been smaller than the F-14 it might have replaced. The attractiveness of the concept was to incorporate 100 percent of the F-22's avionics and 60 percent of the aircraft's subsystems.

The F-4 Phantom was the first-line Navy fighter in the 1960s. The Air Force was forced to take the model, a situation that worked out very well. Early rumors pointed to the possibility that the Navy might have to accept the F-22. That situation appears quite unlikely in the late 1990s. (USAF Photo)

sure might have been greater for this 1990s design time period fighter with the partial dismantling of the defense-industrial concept with economy being the by-word. Congress was not smiling on a multitude of high-buck projects. In fact, those money worries are already cutting into future F-22 plans.

The Lightweight Fighter Competition of the early 1970s had the goal of developing a fighter common to both Air Force and Navy use. Instead, the Air Force selected the single-engine F-16, which the Navy rejected for its use. The Navy then modified the losing YF-17 to the current F/A-18 configuration. The F/A-18 is shown on the left, while the F-16 is on the right. (DOD Photos)

Other Versions of the F-22 27

This artist's concept displays a Navy F-22 Derivative Strike Fighter which was a major modification of the Air Force version. Note the differences in the wings (swing) and cockpit. (Lockheed Photo)

There was even a Navy office established at Wright Patterson Air Force Base, the home of the Air Force's ATF System Program Office at the Aeronautical Systems Division. The Navy group was actually a part of the full NATF management team located at the Naval Air Systems Command in Washington, DC.

But probably the main reason the Navy backed off from all F-22 applications was the worry that adoption of any ATF participation could well jeopardize the on-going F/A-18E/F program which was moving efficiently toward operational service.

After all the Naval applications of the F-22 were finally discarded, the Air Force started looking how different versions of the F-22 could fill other Air Force needs. The service looked at the ATF for replacing such aircraft as the F-15, F-111, and F-117 Stealth Fighter for critical air-to-ground missions.

The extensive Air Force study was headed by the Air Combat Command at Langley Air Force Base, Virginia headed by General Michael Loh. The study looked at the Wild Weasel mission, the destroying of enemy ground-based radars and surface-to-air missiles. Modified F-16s, though, would later be given that important mission, but the F-22 could well field that SEAD (Suppression of Enemy Air Defenses) mission in the future.

With its electronic and offensive capabilities, the F-22 is a natural for this mission. With the plane's weapon load of a pair of GBU-32 1,000 pound Joint Direct Attack Munitions(JDAM), as well as the future Wind Corrected Munitions Dispensers(WCMDs) that are planned for the aircraft.

Loh also indicated that a variant of the F-22 could be designed to take over the mission of the F-111. "The long-range interdiction aircraft, like the F-15, need to be replaced in the future by a stealthy aircraft. There are various options for that, would of which would be a derivative of the F-22," Loh explained.

Other missions that were under consideration, and some of which could come to fruition, include an intelligence mis-

Initially, the F-22 was planned as a replacement for both the F-15 Eagle (Left) and the F-117, however, with the projected fewer numbers of F-22s to be produced, that probably will not be the case. (McDonnell-Douglas and Lockheed Photos)

sion. The prime roles of the aircraft continue to be air superiority and long-range attack, but the sophistication of the plane's electronics enable it to perform many other high-tech missions.

It is assessed that in order to perform that important intelligence mission, it would be necessary to only add a recorder and avionics software improvements to aid in more accurately determining the location of enemy electronic activities. Its potential capabilities could augment the already existing electronic intelligence aircraft, such as the RC-135 Rivet Joint aircraft.

A new near-term weapon system could find its way onto the F-22, a system known as Locas, standing for Low-Cost Autonomous Attack System. Only projected at about 20 inches in length, a Tactical Munitions Dispenser (TMD) could carry as many as six of the weapons. The unpowered vehicles carry an on-board seeker which scans for target identification.

There have also been a number of hypersonic missile designs that could be compatible with the F-22 which could be fitted with the model in the 2005 time period. The weapon, which is currently in an early development stage, could be a follow-on the GBU-28 bunker-busting gravity bomb. The new weapon might incorporate a rocket motor which could accelerate the system to as high as Mach 6 for better target penetration.

Needless to say, there are serious considerations for foreign sales of the F-22 once it enters the expected production phase in about 2003. It remains to be seen whether potential buyers of the plane would get a complete F-22 with all its sophisticated systems, or whether it would be a downgraded version. Such a version would, of course, be cheaper than the full-up F-22, but would it attract any international buyers? That remains to be seen.

It should be recalled that the Air Force earlier had proposed a downgraded version of the F-16 for foreign sales. The plane carried a smaller J79 engine, hence the name F-16/79, but not that first J79 was sold. Could the same situation be faced if downgraded F-22s were to be offered?

At press time (early 1998), there were considerable international feelers out exalting the virtues of the plane, but time will have to tell what the buying response will be. Obviously, there is extremely tough competition right here in the States with the extremely popular F/A-18E/F and improved versions of the F-15 and F-16 providing much lower-cost aircraft with which to contend.

Also, there was the consideration of how the F-22 expenditures would affect the upcoming Joint Strike Fighter (JSF) program which is expected to be in the early development stage in the late 1990s time period.

As with any expensive new system, thoughts of foreign sales certainly are on the minds of both government and contractor officials, and that has certainly been the case with the F-22.

One country which has expressed some interest is Israel which has always been a heavy user of advanced American equipment. The Israeli Air Force currently uses F-15s and would seem to be a logical choice for the F-22.

The Japanese, another user of the F-15, could also be a possible future buyer of the Raptor. But it is possible that if such a purchase were to take place, it could be at least another decade. South Korea could also be considered, but might be a less likely candidate.

The British are also looking into possible alternatives for their Ministry of Defense Future Offensive Air System (FOAS) program. Both the F-22 and JSF systems would have to be considered as potential candidates for that program.

Any European use of the F-22 will also be affected by the consortium-developed Eurofighter and whether it will reach production.

It has been stated, though, by Air Force officials that any F-22 foreign sales would follow US operational use of the plane.

A possible foreign export version of the F-22 has been considered possibly using less sophisticated engines and avionics. It's been unsuccessfully tried before with the F-16 equipped with a less-powerful J79 engine. Not a single sale was made since the customer wanted a plane that could compete with the high-tech aircraft being received by potential adversaries. (Lockheed Photo)

Chapter 5: Parts and Pieces

The F-22 Raptor is 62 feet, one inch long and 16 feet seven inches high. Its wingspan is 44 feet, six inches. These dimensions, appropriately enough, are similar to the fighter's magnificent forerunner, the F-15 Eagle.

Comparisons to the Eagle are inevitable. The Air Force is seeking the same kind of overwhelming air superiority—called "air dominance" during the Raptor's rollout ceremony April 9, 1997—that the Eagle commanded from its inception in the Seventies. However, the Raptor is in a class by itself.

It's made of composite materials as well as aluminum and titanium. Composites make up approximately 50 percent of the total weight of the aircraft. They are used for their strength and weight-saving properties, and they will allow maneuverability at supersonic speeds.

Maneuverability is only one important characteristic, however. The F-22, back in its ATF days, was described as being the result of "the technology of *integrating technologies*." The aircraft would encompass a combination of speed, handling, stealth, integrated avionics and weaponry to give it what the Air Force calls "first-look, first-kill capability," the ability to locate, identify, and destroy an enemy aircraft before its pilot is aware of the F-22's presence.

Its speed comes from two Pratt & Whitney F119-PW-100 engines. The product of seven years of development and two thousand test hours, the augmented F119s—each producing close to an estimated 39,000 pounds of thrust—can propel the Raptor to more than two-and-a-half times the speed of sound.

Evolution of Air Force Air Combat Fighters

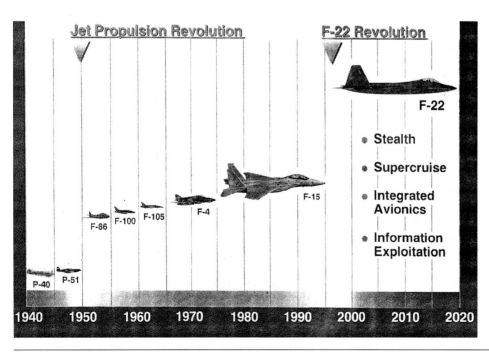

This chart provides the evolution of air combat fighters. (Lockheed-Martin Photo)

The first production aircraft is shown before paint and formal rollout. It would be the lone flight test version until 1998. (Lockheed-Martin Photo)

Some stealth characteristics such as the engine intakes, the leading edges of the wings and vertical tails are evident in this near head-on shot of the YF-22. (Lockheed Photo)

Parts and Pieces

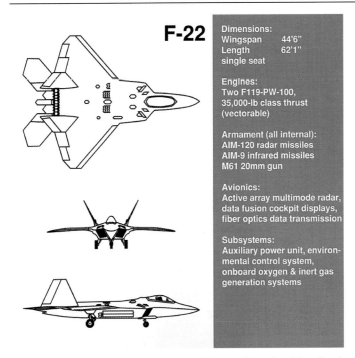

The overall dimensions of the F-22 are given by this chart. (Lockheed Photo)

"THE LAST TIME AMERICA LAUNCHED A NEW AIR SUPERIORITY FIGHTER WAS 7 PRESIDENTS AGO."

Go-Go boots. Love beads. 8-track tapes. These relics have passed into the history books. Yet, we continue to rely on a 30-year-old fighter design that is basically just equal to current foreign models for our national defense. But the F-22 fighter brings a new era. An era of assured air superiority. Dominion through stealth, supercruise, thrust vectoring and advanced avionics. And technology that will allow an F-22 squadron to be supported, maintained and deployed at 30% less cost than current squadrons. F-22. So America will still rule the skies many Presidents from now.

As this advertisement for the F-22 states, it's been a long time since the F-15. (Lockheed Photo)

As important as this brute force is the fact that the F119s also will allow "supercruise," meaning that the F-22 can fly for *relatively* long periods without afterburner. While the Air Force will not specify how long this is, it is a highly desirable trait that will allow the Raptor to cover a lot of ground In a short time without paying the fuel-gulping penalty of augmented thrust using the afterburner.

The F119 is a combination of advanced materials and cooling techniques, state-of-the-art compression aerodynamics—including compressor stage blades which are hollow, and a highly-reliable, full-authority digital engine control, developed by Hamilton-Standard, which coaxes approximately 10,000 pounds more thrust from an engine about the same size as that used in the F-15.

Also worth noting is the fact that the F119 has fewer and more durable parts than older engines and is easier to support and maintain. It uses technology developed under the Advanced Technology Engine Gas Generator (ATEGG), Joint Technology Demonstrator Engine (JTDE), and other engine component research programs funded by the Air Force, Navy, and engine contractors.

Speed and handling also result from airframe shape and weight. The Raptor's weight—more than 60,000 pounds when fully loaded—is comparable to the F-15. The weight is kept relatively low due to extensive use of composites, which make up approximately 56 percent of the aircraft's weight.

Shape is also critical to the aircraft's speed and performance. Its trapezoidal wings are swept back at a 45 degree angle. On them are leading edge slats running from just past the wing root to the tip. Each wing has two flaperons: the one near the wing root is large and rhomboidal; the one towards the tip is small and trapezoidal.

The horizontal tails are canted 60 degrees outward from horizontal and each features full-height rudders making up about one-third of the fin. These knife-thin protrusions are extremely strong and offer little drag even at supersonic speeds.

Two ways of accomplishing stealth. The F-117 in the foreground was the true stealth aircraft. The F-22 will also carry considerable stealthiness. (Lockheed Photo)

32 Lockheed-Martin F-22 Raptor

The new and the old. The F-22 is flanked by an F-15 on the left and an F-16 on the right shown in flight over the California desert. (USAF Photo)

F-22 MATERIAL APPLICATIONS

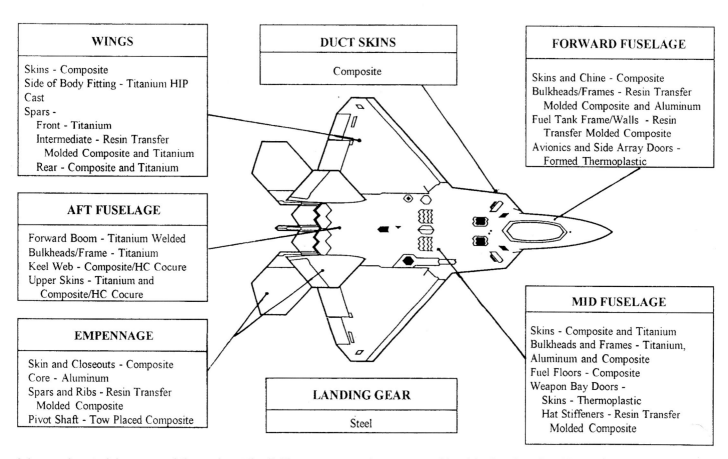

WINGS
Skins - Composite
Side of Body Fitting - Titanium HIP Cast
Spars -
  Front - Titanium
  Intermediate - Resin Transfer Molded Composite and Titanium
  Rear - Composite and Titanium

AFT FUSELAGE
Forward Boom - Titanium Welded
Bulkheads/Frame - Titanium
Keel Web - Composite/HC Cocure
Upper Skins - Titanium and Composite/HC Cocure

EMPENNAGE
Skin and Closeouts - Composite
Core - Aluminum
Spars and Ribs - Resin Transfer Molded Composite
Pivot Shaft - Tow Placed Composite

DUCT SKINS
Composite

LANDING GEAR
Steel

FORWARD FUSELAGE
Skins and Chine - Composite
Bulkheads/Frames - Resin Transfer Molded Composite and Aluminum
Fuel Tank Frame/Walls - Resin Transfer Molded Composite
Avionics and Side Array Doors - Formed Thermoplastic

MID FUSELAGE
Skins - Composite and Titanium
Bulkheads and Frames - Titanium, Aluminum and Composite
Fuel Floors - Composite
Weapon Bay Doors -
  Skins - Thermoplastic
  Hat Stiffeners - Resin Transfer Molded Composite

Advanced materials are used throughout the F-22 structure as demonstrated by this drawing. (Lockheed Martin Photo)

Some fighters push the envelope. This one rips it apart.

Lockheed, Boeing, and General Dynamics have just put the revolutionary YF-22 advanced tactical fighter prototype in the air. This extraordinary aircraft stands as a tribute to the ingenuity and dedication of the employees who designed and built it. These men and women have put America on the path toward the best air-superiority fighter the world has ever known, the F-22.

F-22 ADVANCED TACTICAL FIGHTER
LOCKHEED · BOEING · GENERAL DYNAMICS

Lockheed uses a rainbow to flaunt its newest fighter, the F-22. (Lockheed Advertisement)

AIR TRAFFIC CONTROLLER.

You're looking at the number one air superiority fighter of the 21st century. Faster. Quicker. Stealthier. More reliable and, ultimately, more lethal than any other fighter in the sky.

Designed to replace America's current air superiority fighter, the F-22 has achieved a remarkable balance of capabilities that make it twice as effective as its predecessor. Yet half the cost to operate and maintain.

The F-22 features the most advanced technology ever built into a fighter. A low-observable, composite airframe, revolutionary F119 engines, supercruise (supersonic flight without afterburners), fully integrated avionics and a virtually unlimited angle of attack capability.

All of which means, in the air battles of the future, the F-22 will dominate the skies.

The head-on shot of the F-22 in this advertisement is dominated by the two massive F119 engine intakes which provide the F-22 with a supercruise capability. The ad carries on the theme of "Dominance" in the air battles of the future. (Lockheed Advertisement)

From head-on, the Raptor's nose resembles an inverted triangle whose hypotenuse is a convex curve. This is flanked by imposing rhomboidal engine intakes. The sides of the intakes are flat and taper off near the end of the fuselage at the point where the large, fully-moveable stabilators are positioned. The sleek overall appearance more than suggests speed.

As mentioned before, handling—or maneuverability—is just one component of the Raptor's flying capabilities. Ideally, the aircraft will never be forced to dogfight, but, if the situation arises, the F-22 will be more than capable.

Helping to harness and use the immense power of the aircraft are functions of flight controls, control surfaces, and maybe most importantly, thrust-vectoring nozzles. These nozzles may be directed 20 degrees up or down. When used in conjunction with the large control surfaces, the nozzles allow high angle of attack and choice as to where to point the Raptor's nose.

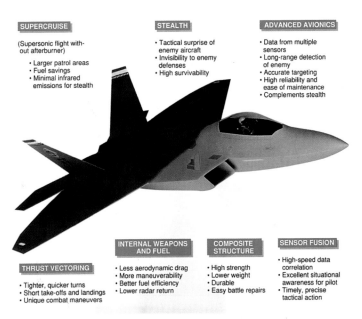

There's a lot new with the F-22. This chart outlines some of the significant technological advancements for the F-22. (Lockheed Photo)

34 Lockheed-Martin F-22 Raptor

The thrust vector control capability of the F119 is shown in this company test. The engine has a plus-or-minus 20 degree capability. (Pratt and Whitney Photo)

The Raptor's basic flight controls are full-digital, fly-by-wire systems—the kind of system pioneered successfully on the YF-16. A flight control computer, developed jointly by General Dynamics and Lear Astronics Corp., features triple-redundancy, which greatly contributes to mission completion and safety.

The flight control system takes input from the pilot and electronically commands the actuators, which move the fighter's control surfaces. Fly-by-wire eliminates conventional mechanical linkage and cables and allows maximum flexibility for tailoring flying qualities.

The flight control system includes three-axis command and stability features which provide precise control and handling. There is also automatic load limiting which allows maximum performance throughout the flight "envelope" while taking into account aircraft loads such as fuel and ordnance.

There is an aileron-rudder interconnect which improves handling even at very high angles-of-attack; the wings' leading-edge flaps are programmed to provide performance-optimum wing camber; and the leading-edge and trailing-edge

A cutaway of the F119 powerplant showing internal components. The Air Force selected this engine in a contract awarded August 1991. (Pratt and Whitney Photo)

The F119 engine is in the 35,000 pound thrust class. It is designed to be more durable and require less maintenance than current fighter engines. (Pratt and Whitney Photo)

Parts and Pieces 35

The winning combination of the F-22 and its F119 engine system. (AEDC Photo)

The F-22 shares some characteristics with current fighters, i.e. the canted twin tails are similar to the F/A-18, while the intake design looks surprisingly like the F-15. (USAF Photo)

flaps all combine for good takeoff and landing characteristics, as well as maneuverability at high speeds.

In addition to flying enhancements, the Raptor's flight control system also offers the following: built-in test for locating faults, minimizing downtime, and completing all pre-flight and system readiness ground checks; VHSIC (Very High Speed Integrated Circuitry)-based computer modules which are common to other onboard systems, thereby reducing the number of different spares and rendering most maintenance tasks as simple as replacing a module; and the capability to make software changes in the computer without removing it from the aircraft. Additionally, all software is programmed in the military standard Ada language.

For the flight control system, the principal subcontractors to the Lockheed-Boeing-General Dynamics team include: for flight control actuators, NML Control Systems, Kalamazoo, Michigan, and Parker Bertes Aerospace, Irvine, California; for the flight control computer, Lear Astonics Corp., Santa Monica, California; for the stick, GEC Avionics, Rochester, England; and for the air data system, Resemont Inc., Aerospace Division, Burnsville, Alabama.

All the handling capabilities of the Raptor are multiplied by its stealth (or low observable) features. Stealth is perhaps the single most important F-22 technology since it contributes to mission accomplishment, safety, and "first-look, first-kill" capability.

For purposes of speed and stealth, there will be a complete absence of external stores. Also note the engine nozzles being exposed such that they can accomplish the thrust vector control function. (Lockheed Photo)

Outstanding all-around vision is achieved with the single piece canopy. (Lockheed Photo)

Parts and Pieces 37

The tricycle landing gear is deployed in this photo illustrating its fairly standard configuration. (AEDC Photo)

The rear control surfaces of the F-22 are shown in a deflected position. The aircraft is probably involved in a ground test in this photo. (USAF Photo)

The F-22's degree of stealth is the result of its shape, engine inlet and exhaust design, and materials. The aircraft's shape—especially its head-on view—shows little or no surface features which might return a radar signal. While the side-view looks more "trackable," one might speculate about Radar Absorbent Material (RAM) coating the sides of the engine compartments.

The inlets—though apparently gaping wide—probably offer an air passageway which winds somewhat to shield the engine fan blades. Likewise, a radar view of the F-22's backside would be just as futile as the head-on look since surfaces perpendicular to the radar beams are so thin.

The Raptor's stealth characteristics are sustained by strictly-internal weapons carriage. The specific weapons suite to be carried by the F-22 is classified as of this writing. How-

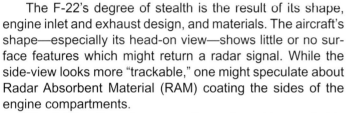

The super-thin stabilizers of the F-22 are outlined in a night sky. (USAF Photo)

The left photo shows an early artist's concept(1990) showing advanced flat panel color displays along with the head-up display. The concept was very close to the final F-22 panel as shown by the simulator on the right. (Lockheed Photo)

Composites were used heavily in the F-22 as illustrated by a part of the wing skin. This is extremely strong and extremely light weight material. (Lockheed Photo)

F-22 Weapons Carriage Capability

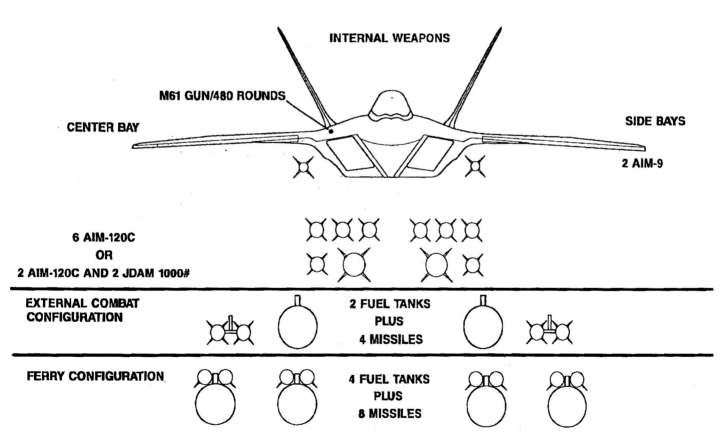

The possible weapon suites for the F-22 are shown by this line drawing. (Lockheed-Martin Photo)

This AMRAAM launch sequence shows the missile falling away from the F-22, then igniting, and accelerating away toward the target. (Lockheed Photo)

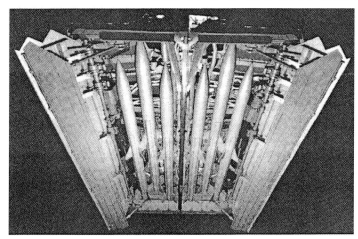

Computer-generated rendering showing a full loadout of six AMRAAMs in the F-22's Main Weapons Bay. (Lockheed-Martin Photo)

ever, one suite could include six AIM-120 AMRAAMs, two AIM-9 Sidewinders (or sophisticated variants) and two 1,000 pound JDAMs. This suite presumably will be enclosed in a remarkable center fuselage underbay. The new fighter also carries a 20mm Gatling Gun similar to that in the F-16.

According to the Air Force's official F-22 fact sheet, the fighter has an "inherent air-to-ground capability, if needed," meaning that the fighter has sensors enabling it to attack surface targets as well as aerial ones.

The Raptor has a powerful radar equipped with an active, electronically-scanned array antenna designed to operate in about 25 modes. This radar makes possible long-range acquisition, tracking and fire control capabilities, and allows the pilot the options of whether to attack from long range or retreat undetected.

The F-22 has radar-detecting equipment and other classified sensors, as well as a datalink through which information is shared with "friendlies" such as AWACS (Airborne Warning And Control System) or others. These work together to pinpoint possible intercept targets.

An AIM-120 is shown being ejected from the weapons bay of the F-22. The missile is physically ejected as opposed to a free-fall launch. (USAF Photo)

If the Raptor's pilot decides to engage, the on-board computers will automatically assign priorities to the targets and pass this information along to the air-to-air weapons. The pilot can move quickly into range and attack before the enemy is aware that he is there.

All of this is portrayed in pictorial representations and symbols on liquid crystal displays which dominate the cockpit. The information displayed quickly conveys information about any type of threat, and the aircraft's computer helps the pilot travel safely through threat environments.

The Raptor's control and display suite consists of: a holographic heads-up display (HUD), one 8x8-inch primary multifunction display (MFD), two 6x6 inch secondary MFDs, two subsystem displays, and an up-front control/display. The primary and secondary MFDs are full-color; the subsystem display and up-front control/display are three-color.

The primary MFD portrays tactical infromation while the left and right secondary MFDs present attack and defensive information respectively. The subsystems displays depict aircraft subsystem and stores management data. The up-front control/display presents communications, navigation and identification information, as well as caution, warning, and advisory data—backed up by voice warnings—and information on corrective actions. The HUD presents primary flight reference, weapon-aiming, and release information.

Data from sensors, threat assessment, and attack management are automated to relieve the pilot of routine tasks and provide situational awareness so that he can maintain a

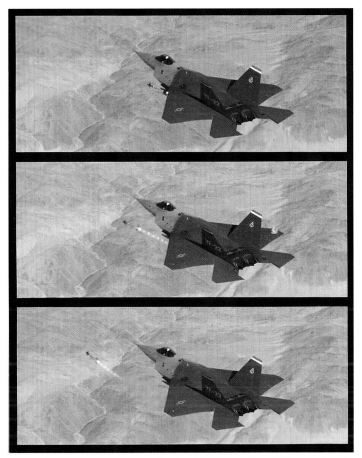

One of the F-22 weapons is the longstanding Sidewinder, in this case an advanced AIM-9M version. (Lockheed Photo)

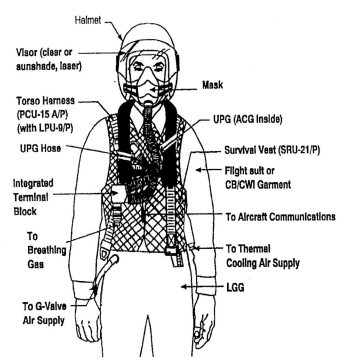

This line drawing of the F-22 Life Support System illustrates the vast complexity of this important component of this fighter. (Lockheed-Martin Photo)

The GBU-32 one-thousand pound class Joint Direct Attack Munition (JDAM) is guided to its target by means of an inertial measurement unit updated in flight with data from global positioning satellites. (Lockheed Photo)

Parts and Pieces 41

Electronic Warfare (EW) Aperture Locations

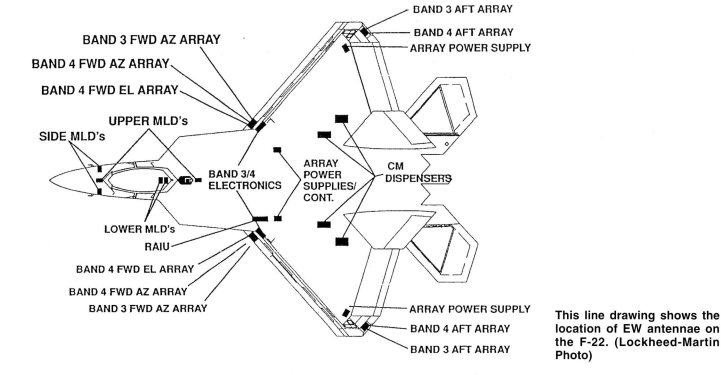

This line drawing shows the location of EW antennae on the F-22. (Lockheed-Martin Photo)

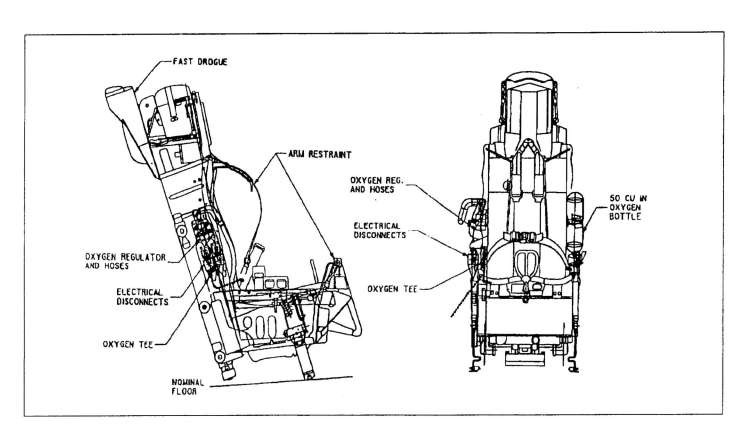

Shown is a line drawing the ACES II ejection seat, the standard seat for current USAF fighters including the F-15, F-16, and, of course, the F-22. (Lockheed-Martin Photo)

huge edge over his adversaries. The result: greater survivability and mission effectiveness than ever before.

The cockpit features a hands-on throttle and stick arrangement which allows the pilot to attack or even dogfight without removing his hands from these controls. The integrated throttle controls both engines, but there are auxiliary controls for independent management of each engine.

The aircraft's life support system affords the pilot greater "g" protection in the form of a fast-reacting "g" valve, positive pressure breathing, and improved "g" suit, which also provides chemical and cold water immersion. The suit is coupled with a personal thermal control system to regulate air temperatures from the environmental control system.

The F-22 uses the tried-and-true, Air Force standard ACES II ejection seat and a canopy jettison system. The Air Force plans to improve the seat by adding leg and arm restraint systems to reduce the chance of injury during ejections taking place above 450 knots.

Other cockpit features include: built-in test from the cockpit, easy component removal and replacement, ejection seat removal without removing the canopy, easy decontamination, HUD replacement without re-boresighting, and the use of components from other Air Force/Department of Defense programs.

The Raptor is designed to accomodate future requirements and new technologies, such as advanced displays and controls, helmet-mounted displays, three-dimensional audio and video displays, and voice controls.

F-22 Integrated Avionics Subsystems

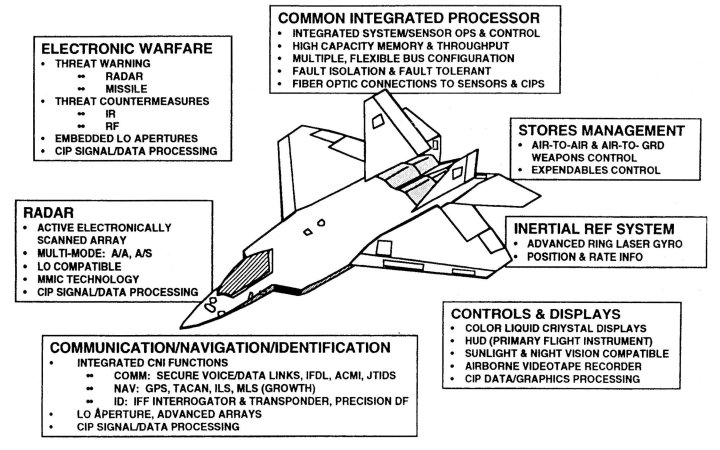

This drawing shows the major components of the F-22's Integrated Avionics Subsystems. (Lockheed-Martin Photo)

Chapter 6:
The F-22 Ground/Flight Test Program

With an aircraft as complex and expensive as the F-22, no stone was left unturned in the testing of the fighter. Every possible aspect of the plane was tested to assure that it would be able to effectively perform its missions. Since it's very unlikely that there will ever be another fighter of this cost, the Air Force wanted to assure that its premier fighter would be up to the job.

Of course, the most visible testing was the flight testing that initially took place with the first two prototypes, the same two planes that had participated in the fly-off competition with the YF-23.

But there was more—much more! Lumped into a broad category that we will call ground testing, the F-22 was tested in wind tunnels in sub-scale form where it was twisted in every attitude, had its stores released, had its engines tested in every conceivable condition, tested the effectiveness of its low-observable signatures, endured live-fire exercises to test battle damage aspects, checked the effectiveness of the F-22's avionics, and on and on.

Let's first look at the important ground activities which will pay big dividends when the plane reaches operational status.

Ground Testing

With the weird new shapes of modern aircraft, there is always concern of exactly what will happen when ordnance is dropped from the plane. That was a particular concern with the F-22, with its capability to carry numerous missiles and stores and support a wide variety of loadings and downloadings on four wing stations, two side bay stations, and six main bay stations.

At the Air Force's Arnold Engineering Development Center, F-22 stores separation tests were carried out since 1992 at various angles of attack and stores orientations. All types of data was acquired to monitor the drop procedures. The testing capabilities of the 4T and 16T tunnels at AEDC enabled the testing of missile clusters, the capability to study a tumbling store, and the mechanics of releasing wing-mounted fuel tanks.

A wind tunnel model of the F-22 ready for testing in the 16 foot transonic wind tunnel at the Arnold Engineering Development Center in 1992. (USAF Photo)

An F-22 model, created by an Air Force Super Computer, as part of a technology called computational fluid dynamics which is used to predict airflow over the aircraft surface. (USAF Photo)

The red areas on the computer image represent area of higher air flow. The areas of highest airflow occur over the cockpit and the rear vertical stabilizers. (USAF Photo)

The Air Force used computational fluid dynamics to decrease the amount of wind tunnel tests, which are extremely expensive. (USAF Photo)

The testing continued through 1997, and greatly reduced the risks associated with the actual releases in flight.

Of course, the testing of the aerodynamics of the plane itself were of paramount performance. In the 16T tunnel, models of the F-22 were tested at various angles of attack and Mach numbers to determine flow characteristics. The resulting aerodynamic forces and moments helped predict how the actual aircraft will perform as throttle settings vary.

Other testing at AEDC helped determine the effects of the engine nozzle pressure ratio and nozzle geometry the F-22's aerodynamic characteristics. Test conditions were varied on the model to simulate the flight conditions the F-22 would normally encounter. Mach number was varied from 0.40 to 1.50. The aircraft angle of attack was varied from minus 2 to 84 degrees, while angle of sideslip was varied from minus 6 to 6 degrees.

The jet effects test article was a 1/11-scale of the F-22. The fully-metric model with faired intakes was supported by a hockey-stick shaped strut to minimize interference on the afterbody. The model was also equipped with remotely-actuated rudders and horizontal tails capable off plus-or-minus 6 degree deflections.

Wind tunnel testing of the F-22's Advanced Infrared Search and Track (IRST) system were also accomplished at the AEDC facility. The system was mounted on a full-scale model of an F-22 aircraft forebody in the facility's 16-foot transonic tunnel.

The purpose of the testing was to validate aero-optical performance and verify the aero-thermal environment of the IRST system. The IRST system was embedded in the F-22 chin section to retain the supersonic cruise and low-observable characteristics of the aircraft. Optical sources were also

A F-22 wind tunnel model is maneuvered to a very high angle of attack in an AEDC wind tunnel in 1995.

The F-22 Ground/Flight Test Program 45

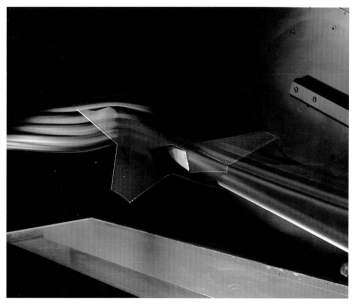

To envision airflow patterns around this F-22 model, Lockheed engineers injected smoke into a wind tunnel and illuminated it with a laser. By late 1990, Lockheed had completed 24,000 hours of wind tunnel testing on the F-22. (Lockheed Photo)

This Lockheed wind tunnel model was used to test the spin characteristics of the F-22. It's one/twelfth scale. (Lockheed Photo)

mounted at various areas in the tunnel where they could be viewed by the sensor.

The purpose of the IRST system is to detect two types of hot emissions from threat aircraft-plume emissions and the leading edges that become hot due to the movement of the aircraft through the air.

The F119 engine has also been put through its paces since the late 1980s. The General Electric F120 engine was also certified during the testing, but once the F119 was selected as the winner, the testing continued on it alone.

The testing involved evaluating the engines aeromechanical performance, combustor and augmentor

The P&W F119 engine during test. Actual thrust is thought to be between 35,000 and 39,000 pounds. (P&W Photo)

Shown here is the F119 F-22 engine under test at the Arnold Engineering Development Center ASTF wind tunnel facility. (USAF Photo)

The engine competition, in the form of the General Electric F120, is shown in ground test illustrating its thrust vectoring capability. (General Electric Photo)

operability, vectored and non-vectored nozzle performance, fan performance, compressor stall margin and airstart capability.

The test team was successful in obtaining engine data at the highest planned engine inlet pressure ever tested. More aeromechanical data was processed on this test project than any other engine previously tested at AEDC.

There were several new subsystems requested by Pratt & Whitney including a supplemental engine cooling system and an increased capacity digital data system used to obtain critical engine temperatures. Other support systems supplied by AEDC for F119 testing included a high-response pressure sensor support system, an automatic nitrogen purge system, and a cold-fuel conditioning system.

Pratt & Whitney also accomplished endurance qualification testing in 1996 on the F119. The test engine endured a thousand cycles, simulating actual combat missions, at the P&W facility at West Palm Beach, Florida.

The Air Force's F-22 System Program Office from Wright Patterson Air Force Base accomplished a series of live fire testing on the F-22's wing structure at Boeing's Tulalip Test Site in Seattle, Washington.

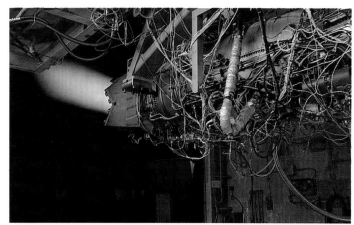
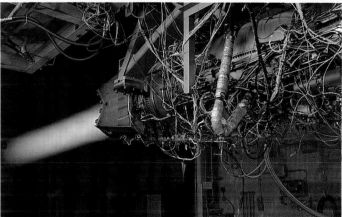
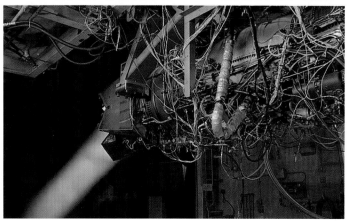

The F119's thrust vector control, shown here under test, is capable of plus-or-minus 20 degrees. (P&W Photo)

The testing was started in 1992 and involved shooting a ballistic weapon at a wing box test article, to simulate combat gunfire and its effects.

Since some of the new materials on the F-22 had not been exposed to combat conditions, simulated combat testing allowed the new design to be tested without the loss of life or equipment. The results of the testing paid major dividends in guiding modifications to the wing design in order to enhance its vulnerability posture.

A large caliber projectile blasts through an F-22 wing section. A mirror at the bottom right displays a reverse angle view of the impact. Live fire tests allow engineers to simulate gunfire effects on new materials and construction techniques. (USAF Photo)

One of the characteristics of the F-22 is its use of advanced materials. Here, a Lockheed technician in 1990 tests the strength of a thermo-plastic resin composite with laser holography, a version of the materials which would eventually find their way to the F-22. (Lockheed Photo)

An F-22 wing skin panel is seen here undergoing through-transmission ultrasonic testing at Boeing. This testing provides a non-destructive means to insure the manufacturing quality of the skin built from advanced composite materials. (Boeing Photo)

A calibration chamber at Lockheed's research and development center in California was used in the early 1990s to verify low radar signatures of antennae mounted on models of the YF-22. (Lockheed Photo)

This view shows one of three test chambers at the Lockheed calibration facility. It helps measure electronic emmissions from the aircraft which could identify it to enemy radars. (Lockheed Photo)

A Boeing electronics technician tests a circuit board in one of the simulation racks in the company's F-22 Avionics Integration Laboratory in Seattle. Software engineers planned to use the lab to develop and test the F-22's avionics. (Boeing Photo)

LEFT: This Lockheed F-22 simulator allowed test pilots to check out the characteristics of the YF-22. This is one of two domed F-22 simulators, this particular one being located at Kelly Johnson Research and Development Center. (Lockheed Photo) RIGHT: On the screens are computer-generated images of fighter aircraft being tested in the Lockheed F-22 simulator. (Lockheed Photo)

A pilot enters the 28-foot high simulation dome used during the early stages of the F-22 program. (Lockheed Photo)

The Interactive Control Station, at Lockheed's Kelly Johnson facility, portrays threat assessment data similar to that is found on modern F-22s. (Lockheed Photo)

Northrop Grumman also got into the F-22 test game with 18 months of system-level integration and testing on the developmental radar. The AN/APG-77 radar was the first of 11 systems to be delivered during the engineering manufacturing development (EMD) phase of the F-22 program.

The Air Force also conducted extensive testing to assure that there would be no risk with the F-22's antenna system.

The testing involved the use of the YF-22 which had crashed in 1992. Since about 25 percent of the structure had been damaged, the left wing, tail and stabilizer were recreated in wood to simulate the plane's shape. The testing was carried out at the Air Force's Rome(New York) Laboratory.

The test plane was mounted on a three-axis positioner which was capable of handling the 13-ton aircraft. The tests

Submerged in 42 degree water for two hours, this volunteer was testing an advanced protective flight suit. The suit was designed to protect F-22 pilots from hypothermia should they have to eject over cold water. (Boeing Photo)

One of the candidates for the F-22 helmet is shown here. The HGU-86/P helmet incorporated lighter weight, stronger, and more resistant to impact than previous helmets. (USAF Photo)

This Boeing 757 test aircraft was used to test YF-22 avionics. It is called the 'Avionics Flying Lab'. The equipment allowed engineers to monitor and evaluate the YF-22's avionics performance, trouble-shoot problems, and develop remedies for modifications in a flight environment complementing ground based testing. (Boeing Photo)

involved the rotating and elevating of the F-22 to allow the incoming RF signals to arrive at different angles to the plane's antennas.

Almost every aircraft attitude was available where the nose could be lifted or rolled plus-or-minus 60 degrees and can be rotated a full 360 degrees.

Other antenna testing for the F-22 was also accomplished by Lockheed Sanders which used F-22 scale models in indoor company ranges. The chambers were lined with radar-absorbing materials allowing the capability to gather data on antenna patterns and look angles.

Before the first flight of the YF-22, the Air Force at Kirtland Air Force Base used simulators to evaluate the plane's performance against current and future threats. The F-22's closely-coupled avionics system required the testing as an entire system, avoiding separate assessments of other subsystems.

This artist's concept shows a proposed test of the F-22 wing structure and nose section mounted on the front of the transport. (Boeing Photo)

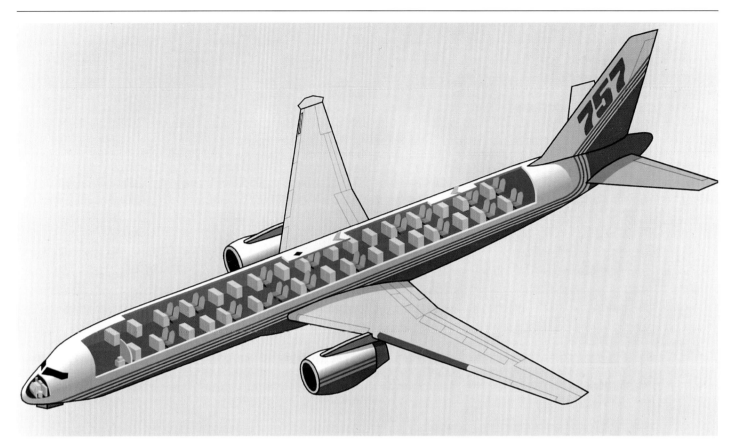
This cutaway of the 757 test aircraft shows multiple test monitoring locations in support of the YF-22 test program. (Boeing Photo)

A view of the YF-22 equipped with a special spin recover chute apparatus during a series of test flights. These tests involved flying at speeds as slow as 80 knots and angles of attack as high as 60 degrees. (Lockheed Photo)

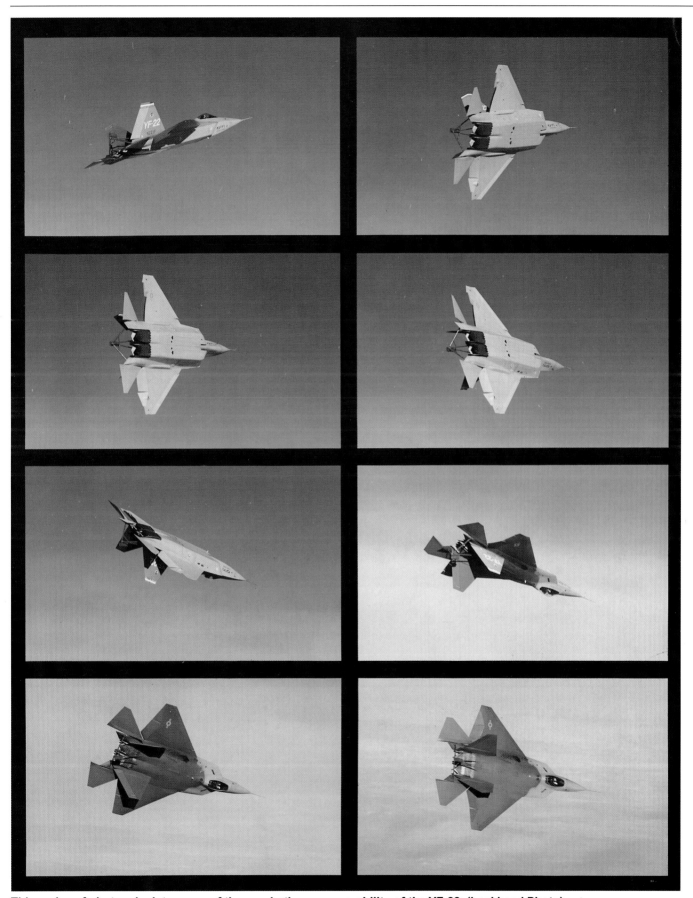
This series of photos depicts some of the aerobatic maneuverability of the YF-22. (Lockheed Photo)

The pair of YF-22 prototypes are shown flying in formation prior to the ATF selection. The aircraft carried both the F119 and F120 engines. (USAF Photo)

Lockheed used another simulator, the Vehicle System Simulator (VSS) to test flight-critical systems before the first test flight. The Fort Worth-based facility tested the F-22 hydraulic and electrical systems shortly before the first prototype took to the air. The locations of the various components are very close to their relative location on the real aircraft. All the routing between the components was also closely duplicated.

The VSS actually reduces the requirements of the flight test program by accomplishing tests which would be dangerous to attempt on a flying aircraft, serious situations such as double engine shutdowns or control system failures.

A new Avionics Integration Laboratory has also been built by Boeing to perform final ground testing of the avionics systems.

Since the F-22 could well be operating in a nuclear environment, scientists at the Lawrence Livermore National Labo-

One of the YF-22s is flaring out for landing with F-15 chase plane in the background. (USAF Photo)

The YF-22 demonstrates a nose-high attitude during this refueling from a KC-135 tanker. (Lockheed Photo)

Refueling was part of the ATF selection process criteria as is shown here. Note the location of the refueling receptacle located aft of the cockpit. (Lockheed Photo)

ratory developed nuclear computer simulations to determine the effects on the plane. A computer program called the DYN3D (for Dynamics in Three Dimensions) predicted when the F-22 might bend and bounce back in certain nuclear scenarios.

A more likely danger to the F-22, though, is actual battle damage and Boeing, at its Tulalip, Washington facility actually fired live 30 mm incendiary ammunition at its manufactured component, the aft fuselage section to determine its durability.

The test results were very close to those predicted by modeling techniques. In addition to the aft fuselage section, Boeing also accomplished live fire tests on both wing sections and aft booms.

Air Force pilots of the future also have a better chance of surviving emergency ejection over cold water, thanks to a new protective suit designed by Boeing and META Research, Inc.

In cold-water tests conducted in 1995, the body temperature of volunteer test subjects wearing the suit fell no more than a fraction of a degree after the volunteers spent two hours submerged in water measuring a cool 45 degrees F. The tests exceeded Air Force requirements that body temperature not drop below two degrees after two hours.

A feature unique to the suit is the capability for the pilot to adjust the temperature and flow of air to the inside of the suit while in flight to achieve body cooling. A cooling line in the suit distributes cool air over the entire body, minimizing the loss of body fluid through perspiration.

The effects of vibration were also tested on the F-22 using a unique test device. Lockheed and Air Force engineers certified that the Raptor would be able to fly at the speeds and altitudes prescribed for the first test flight.

This YF-22, carrying the eventual-winning F119 P&W engine, benefited from the thrust vectoring control capability of this engine. (Lockheed Photo)

The F-22 Ground/Flight Test Program 55

One of the blackest days of the early F-22 flight test program was an landing accident when the plane encountered the runway and was burned on its port side wing and tail. The plane was not rebuilt and flight testing did not resume until 1997 with the first production line model. (Lockheed Photo)

See photocopy in order to generate caption.

Historically, airframes have been excited by using external electrodynamic shakers which vibrate the airframe. With the F-22, vibrations were produced by using onboard flight control systems to oscillate the control surfaces. Understanding the effects of airframe vibration is important to the high-performance F-22 because of its dynamic flight environment.

Airborne Testing

With aid of a modified Boeing 757 transport, the capability of the F-22 avionics system was tested for its ability to acquire and integrate data from a variety of aircraft sensors and effectively display it on advanced cockpit displays.

The F-22 electronics and software that made up the plane's avionics system began testing in 1990 aboard the Avionics Flying Laboratory (AFL) after a successful ground demonstration of the avionics system.

Avionics system sensors flying aboard the AFL included the Active Array Radar(built by Westinghouse and Texas Instruments), Communications-Navigation Identification(built by a team led by TRW), Electronic Combat(built by Lockheed and General Electric) and the Infrared Search-and-Track System(built by General Electric).

Flight Testing (Pre-ATF Selection)

Of course, the initial flight testing for the F-22 began when it was in competition with the YF-23 for actual selection.

Its first flight was accomplished on September 29, 1990 flying from the Lockheed production facility to Edwards Air Force Base. The second YF-22 flew for the first time shortly thereafter.

During the evaluation period, significant testing of the aircraft occurred including low-speed taxi tests, the engaging of the thrust-vectoring nozzles, and demonstrated airspeeds as low as 120 knots and roll rates in excess of 100 degrees per second.

In November of 1990, the pair of prototypes had achieved a significant period of supersonic flight, including 22 minutes on a single flight.

The first production F-22 takes to the air at the Lockheed Martin facility on September 8, 1997. (USAF Photo)

There were also successful refueling tests accomplished during this period. Some concern existed for this critical operation since the F-22 fuel receptacle is located far back on the upper fuselage causing the refueling boom to come very close to the canopy.

A huge milestone was reached in November of 1990 when the first air-to-air missile(an AIM-9M Sidewinder) was launched for the first time from the Number Two YF-22 prototype. The Air Force indicated at the time that the accomplishment was significant because the prototype was so close to the production version.

The YF-22 prototypes also demonstrated one of their prime design attributes with excellent high angle of attack flight.

AOAs as high as 60 degrees were achieved. All of the high AOA testing was accomplished by the number one prototype, which was powered by the losing GE F120 engines.

For this round of testing, the prototype was equipped with a special apparatus containing a stabilization recovery chute, to be deployed if the aircraft entered an uncontrollable spin. After a test deployment in December 1990 to confirm its operation, the chute was never used again.

There were several minor glitches that plagued the competition flight testing with a single in-flight engine shutdown and the failure to raise the landing gear due to computer problems.

Flight Testing (Post-ATF Selection)

In 1991, the F-22 selection was made and flight testing continued with the pair of prototypes. Of course, the prototype carrying the General Electric engine saw its powerplants substituted with the winning Pratt & Whitney engines.

Then the problems began with the testing at Edwards Air Force Base. It was the worse thing that could happen to an expensive, and somewhat controversial program, when one of the prototypes crashed during a landing attempt.

As the plane neared the runway, it suddenly started pitching up and down before slamming into the runway. It then skidded a considerable distance before catching fire. The flight was being used to gather additional data to help define the design before the test versions were built.

The crash prompted a suspension of additional flights. The plane was too badly damaged to be repaired, and with the second prototype located at Lockheed for ground testing, there was no additional flight testing accomplished until the first production flight test vehicles became ready in 1997. Five years with no flight testing. It was not a good situation, but one that the Air Force had to live with.

Flight Testing (Pre-Production Vehicles)

In April 1997, the pomp and circumstance came on strong with the dedication of the first production Raptor. From that time on, the call for cost reduction would follow the program.

That first model, Number 4001, dubbed Raptor 01, was of course, the star of the show with patriotic paint adorning its sheet metal. But immediately following the ceremony the plane was stripped of its formal paint and prepared for its first flight test.

A fuel tank leak was located in the F-1A tank just aft of the cockpit canopy. The extremely minor leak only dripped at about one drop every six seconds, but it had to be repaired before the first flight. The oil problem had developed in the APU/auxiliary generator area. The worry occurred when the oil pressure started dropping when the oil temperature climbed. That problem was also repaired before the first flight. There were also landing gear problems that had to be repaired. Initially planned for June of 1997, fuel tank leakage and oil problems delayed the first flight until September 8.

Lockheed Martin Chief Test Pilot Paul Metz was at the wheel with the Raptor's powerplants operating at military power. In the first-flight profile, the plane reached an altitude of 15,000 feet. The first flight would be the only flight at the contractor facility. Then, the wings were removed and the plane was ferried to Edwards Air Force Base, California where the flight test program was initiated.

It could well be the most important flight test program in US Air Force history.

The fancy detailing that was in place for the dedication had been removed for the flight test program. Note the Raptor 01 in red lettering on the tail. (USAF Photo)

Chapter 7:
F-22 Production and Operational Service

The F-22 is an expensive aircraft, make no mistake about that. And for that reason, the Air Force looked for ways to keep the production costs as low as possible.

Derived by the F-22 System Program Office, along with all the industry partners, a lean manufacturing plan coined the Engineering Manufacturing Development and Production Program was instituted. The goal of the plan, which was introduced in 1994, was to produce a 40-60 percent reduction in total production time.

The time-saving was to be accomplished streamlining major processes by eliminating all forms of waste and low-value activity. A number of F-22 participants actually visited the Toyota Production Facility in Georgetown, Kentucky to learn first hand about the successful production techniques of that car company.

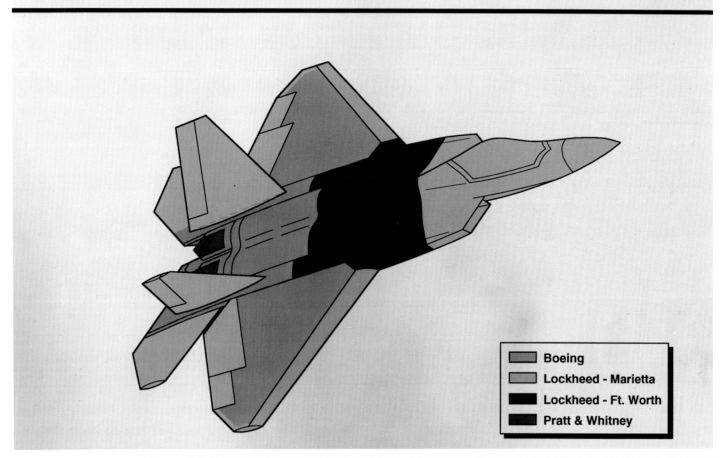

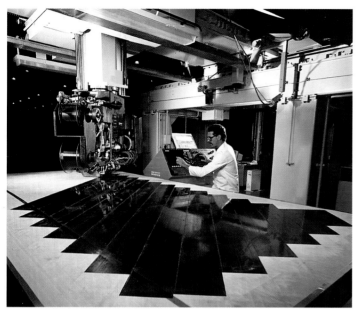

A Lockheed technician cuts strips of thermoplastic-resin composites for an aircraft panel. These composites are at least as strong as their metal counterparts and often are easier to repair. The F-22 is taking advantage of this technology. (Lockheed-Martin Photo)

Giant autoclaves at Lockheed cure advanced composite materials used on the F-22. (Lockheed-Martin Photo)

To accommodate the production flight test portion of the program, seven single-seat F-22As and two tandem-seat F-22Bs were to be produced. Two ground-test articles also were to be constructed.

Initially, there were four pre-production verification aircraft planned, but that number was dropped to two in 1996 for cost reasons.

The production figures given for the program initially called for 438 aircraft, but when the complete program is completed, many felt that number could be considerably reduced. That projection could come to pass if the results of the Quadrennial Defense Review (QDR) in 1997 are accepted.

The Review recommended a massive cut of 99 Raptors which dropped the wing equivalents from 20 to 19. That would also reduce the final production total to 339 aircraft. Of course, this was not a final decision, but it was thought to be the direction the production program would follow. The possible long-term effects of this reduction could be possible reduction of production rates near the end of the program and raise the unit cost of the fighter.

Interestingly, the competitive F-18E/F Navy program received an even-greater reduction, with the projected thousand of the model being significantly reduced to 548 aircraft.

The point was made, however, in the case of both the F-22 and F-18E/F, the possibility of derivative versions of both models could emerge. If any such F-22 versions should evolve, they probably would not be included in the 339 total. It was stated that such derivatives could occur in the 2012 time period and possibly serve as a replacement for the F-15E/F-117 models.

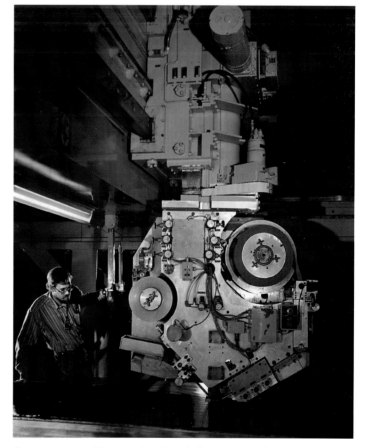

A process used to construct parts is automated composite tape laying shown here in the former General Dynamics plant in Fort Worth, Texas. (General Dynamics Photo)

F-22 Production and Operational Service 59

F-22 forward fuselage being moved into the mate position. (Photo courtesy Lockheed-Martin)

Workers at Lockheed-Martin guide a large titanium bulkhead into an assembly fixture for the F-22 mid-fuselage. Lockheed-Martin Photo)

Boeing machinists move the titanium and composite rear fuselage for the F-22 into position for high-precision automated drilling. A data-driven laser-guided drilling machine will drill more than 2,000 holes based on engineering data fed into a computer. The holes are needed for upper composite skin and lower engine-bay door attachments. (Boeing Photo)

Boeing crane operators load the aft fuselage for the first F-22 fighter into a shipping container at the Boeing Development Center in Seattle. The part was flown to Marietta, Georgia aboard a C-5 military cargo transport. (Boeing Photo)

The aft fuselage houses the two Pratt and Whitney F119 engines that will power the stealthy fighter. It also carries fuel and supports the wing and tail surfaces. When completed, the aft fuselage will measure about 19 feet long by 12 feet wide. (Boeing Photo)

Boeing crane operators and assembly mechanics load a center keel into an assembly fixture. The installation is one of the first steps in starting major assembly of the aft fuselage for the F-22. (Boeing Photo)

The left-hand wing of the first F-22 is lowered onto a pallet before being delivered to Lockheed-Martin in Marietta, Georgia. Boeing delivered the first set of wings on November 8, 1996. Boeing builds the titanium and composite wings in Seattle using advanced production techniques such as resin transfer molding and automated precision drilling. (Boeing Photo)

Boeing employees operate and monitor an automated tape-laying machine that applies layers of advanced composite resin-impregnated tape atop a tool to form the wing part. (Boeing Photo)

F-22 Production and Operational Service 61

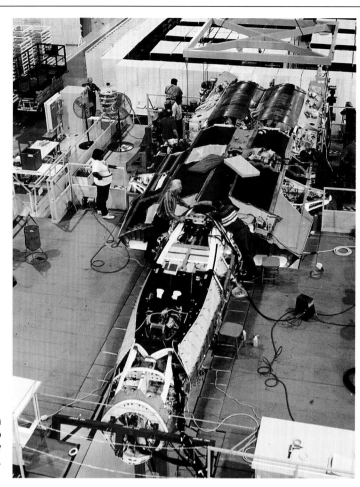

The aft fuselage is shown in this photo being lowered into position. It completes primary assembly of the F-22 fuselage. (Lockheed-Martin Photo)

The assembly jig for the F-22 wing is shown at the Boeing, Seattle facility. (Boeing Photo)

The following is the proposed outline of F-22 production(as of mid-1997) without the proposed reductions. It would be some time before all these implications would be sorted out.

Mid-1999— First flight of an F-22 with a full avionics suite
Early 2000— Lot 2 contract for six F-22s for 2002 delivery
Early 2001— Lot 3 contract for 12 F-22s for 2003 delivery
Early 2003— Lot 4 contract for 20 F-22s for 2003/4 delivery
Early 2004— Lot 5 contract for 30 F-22s for 2004/5 delivery
Late 2004— Initial Operational Capability
2013— Delivery of last planned production F-22 (depending on production schedule finalization.

Production Responsibilities

The F-22 can be divided into four main pieces from a manufacturing responsibilities aspect. Prime contractor Lockheed Martin is responsible for the forward and mid-fuselage. Boeing's responsibilities lie with the aft-fuselage and wings. The final assembly of all the pieces is accomplished at Lockheed Martin's Marietta, Georgia facility.

Following is a discussion of the manufacturing aspects of these parts:

Forward Fuselage and Empennage

The Forward Fuselage consists of the structure aft of the radar bulkhead, the cockpit area, the front radome, nose wheel well, and F-1 fuel tank. It consists of about 3,000 parts made mostly of aluminum and composite materials.

The Forward Fuselage also contains wiring harnesses, tubing, cockpit instrument fixtures, avionics racks, and canopy mounts. The structure is just only 17 feet long, slightly wider than five feet, and five feet, eight inches tall, and weighs roughly 1,700 pounds.

Built up in two sections, the Forward Fuselage is joined together by two long and relatively-wide side beams and two longerons that run the length of the assembly. The canopy, which is built by Lockheed Martin, is 140 inches long, 27 inches high, and weighs about 350 pounds.

In the production process, this assembly mates directly to the Mid-Fuselage Section.

F-22 Final Assembly
Lockheed Martin Aeronautical Systems, Marietta, GA

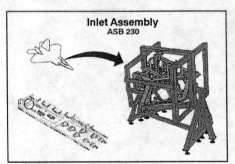
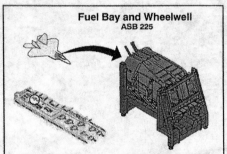
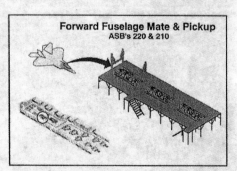
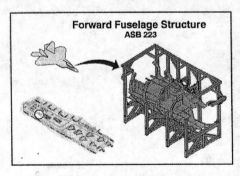
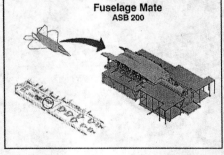
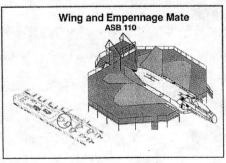
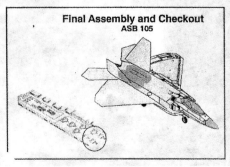

A 1995 Lockheed national advertisement tries to kick up excitement for the new fighter. (Lockheed-Martin Advertisement)

Like all the previous fighters, the F-22 will possess a refueling capability to 'stretch its legs.' (Air Force Photo)

Mid-Fuselage Section
Lockheed-Martin constructs the Mid-Fuselage section of the aircraft at its Tactical Aircraft Systems in Fort Worth, Texas. The final assembly operations for the F-22 is being assembled at the prime contractor's Marietta, Georgia facility.

The Mid-Fuselage Section is the largest and most complex of the F-22 sections being 17 feet long, 15 feet wide, and six feet high. It weighs in at 8,500 pounds. This section is considered the heart of the F-22 as almost all systems pass through this section, including the hydraulic, electrical, environmental control, and auxiliary power systems as well as the aircraft's fuel supply. In addition, there are three fuel tanks, four internal weapons bays, the 20 mm cannon, and the auxiliary power unit(APU).

Lockheed Martin uses a modular approach to assemble the mid-fuselage. Three modules, which are simultaneously assembled prior to mating, make up the complete structure. Because of its width, the Mid-Fuselage has to sit at almost a 45-degree angle in its reusable metal shipping container. This is so its shipping box will fit on a flatbed truck and still be allowed on roads from Texas to Georgia.

Aft-Fuselage Section and Wings
Boeing develops and constructs the wings, the Aft Fuselage,

This special F-22 maintenance cart makes it easy to maintain the fighter's F119 engines. Engine removal and reinstallation time is estimated to be about 40 percent faster than with current equipment. In this photo, Boeing and Air Force technicians are preparing the trailer for the engine's removal from the aft section. The cart uses no hydraulics because hydraulic fluid is a hazardous material and leaks can cost expensive delays. (Boeing Photo)

Major Greg Neubeck, F-22 Action Officer from Langley Air Force Base, Virginia 'flies' a cockpit development simulator to sample Boeing-developed F-22 head-down tactical displays. Boeing designer Marshall Williams gathers feedback. (Boeing Photo)

along with installing the engines, nozzles, and auxiliary power unit. Boeing is also integrating and demonstrating prototype functionality of the radar and the infrared search and track system (IRST).

The Aft Fuselage Section houses the two P&W engines along with containing the environmental control system, fuel, electrical, hydraulic, and engine subsystems.

Approximately 25 percent of the Aft Fuselage is comprised of large electron beam-welded titanium forward and aft booms. The largest of these booms, the forward boom, is more than ten feet long and weighs approximately 650 pounds. The welded booms of the aft fuselage are extremely weight-efficient and reduce the use of traditional fasteners buy approximately 75 percent.

The Aft Fuselage is shipped to Marietta is shipped to Marietta in a reusable metal container that fits upright in a rail car, or can be placed on its side for shipping by cargo aircraft. The first aft fuselage was delivered to Marietta aboard a C-5 Galaxy transport aircraft.

Operational Status

The F-22 is expected to become operational in the 2003 time period, but as was the case with the production total, this time period is subject to change, and change again. Such is the case of high-cost military systems in the 1990s.

During 1997, an interesting concept for the deployment of the F-22 was proposed. Called the Vanguard Force, the F-22 would be a part of a tri-service quick-reaction force that could be quickly deployed anywhere in the world.

The Vanguard Force would prove the effectiveness of the F-22 before large-scale production would be approved. Of course, there was objection from the services who felt that

In this artist's rendering, a Lockheed-Martin-Boeing F-22 Air Dominance fighter has destroyed one of two adversary aircraft and has launched an AMRAAM at the remaining enemy fighter. It is shown in a variation of the camouflage scheme currently used on the F-15. This paint scheme is one of several that the Air Force evaluated for use on operational F-22s. (Lockheed-Martin Photo)

each would lose itsr identity. Time will determine if this concept ever comes to pass.

But be assured, the Air Force really wants the F-22, and it wants them in sizable numbers. Studies have shown that the F-22 could have as much as triple the effectiveness of the F-15 in some areas. That's certainly an impressive credential considering that the F-15 is still considered the best fighter in the world in the late 1990s.

Shown in an artist's concept is an F-22 taking off on a training mission from Tyndall Air Force Base, Florida. The Air Force announced that the 325th Fighter Wing at Tyndall will be the training wing for F-22 pilots. (Lockheed-Martin Photo)

This artist's concept clearly shows a possible camouflage scheme for the F-22 which had not been determined by presstime. (Lockheed-Martin Photo)

F-22 Production and Operational Service

COLOR GALLERY

All photos courtesy of Lockheed-Martin unless otherwise specified. (above: Photo Courtesy Northrop)

70   Lockheed-Martin F-22 Raptor

F119-PW-100 Turbofan

Engine Characteristics

Type:	Twin-Spool Augmented Turbofan
Application:	F-22 Advanced Tactical Fighter
Thrust:	35,000 Pound Class
Engine Control:	Full Authority Digital Electronic Control
Compression System:	Twin-Spool/Counter Rotating/Axial Flow/Low Aspect Ratio
	• 3 Stage Fan
	• 6 Stage Compressor
Combustor:	Annular
Turbine:	Axial Flow/Counter Rotating
	• 1 Stage High Pressure Turbine
	• 1 Stage Low Pressure Turbine
Nozzle:	Two Dimensional Vectoring Convergent/Divergent

Milestones

First Sea Level Test	December 1988
First Flight	August 1990
Selected for the F-22	April 1991
Contract Award	August 1991
First Production Engine	Mid 1998

(Photo Courtesy Pratt & Whitney)

72 Lockheed-Martin F-22 Raptor